SPACE

CHELSEA HOUSE
PUBLISHERS
A Haights Cross Communications Company ®
Philadelphia

First hardcover library edition published
in the United States of America in
2006 by Chelsea House Publishers,
a subsidiary of Haights Cross Communications.
All rights reserved.

A Haights Cross Communications ✦ Company ®

www.chelseahouse.com

Library of Congress Cataloging-in-Publication
applied for.
ISBN 0-7910-9009-4

Project and realization
Parramón, Inc.

Texts
Eduardo Banquieri

Translator
Patrick Clark

Graphic Design and Typesetting
Toni Inglés Studio

Illustrations
Marcel Socías Studio

First edition - March 2005

Printed in Spain
© Parramón Ediciones, S.A. – 2005
Ronda de Sant Pere, 5, 4ª planta
08010 Barcelona (España)
Norma Editorial Group

www.parramon.com

BEYOND THE INFINITE

To explore our immense universe in just a few pages would be impossible. Instead, this book describes some of the main characteristics of our universe—its origins and the elements of which it is composed; some of its most striking phenomena (comets, shooting stars, meteorites, and eclipses); and some of the methods and instruments used to study it (telescopes and spacecraft).

We hope this book will inspire passion for the study of the universe and encourage young people to read more in-depth works on the subject.

SPACE ODYSSEY

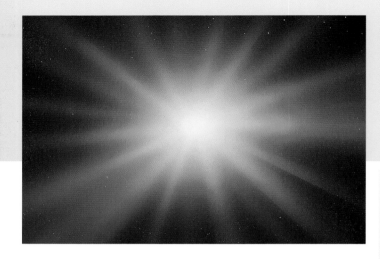

It is believed by many that nearly 4.6 billion years ago, a large explosion occurred. It is possible that the first particles that would later form the universe were created in just fractions of a second.

THE MYSTERIOUS UNIVERSE

The universe has always been a mystery to human beings—so much so that since ancient times, people have adored and treated the planets and stars like gods. The sun represented God for the people of many ancient civilizations. They believed the giant star was the creator, who gave life and took it away. Meteorites, shooting stars, and eclipses were seen as divine signs. To witness such occurrences meant something good or bad could be on the way. In some Polynesian cultures, certain phenomena, such as a new moon or an eclipse, were viewed as an expression of the anger of the gods, who made the moon or sun disappear to punish human beings. People believed they had to carry out a human sacrifice to make the sun or moon appear again. Thus began the adoration of the mysterious universe. Beliefs like these that revolved around respect for the power of the universe dominated human life for many centuries.

ASTRONOMY IN ANTIQUITY

The most technically and scientifically advanced early civilizations (such as Babylon, Egypt, India, Phoenicia, China, and the Mayan culture of Central America) were especially interested in better understanding the universe and its phenomena. Astronomy developed out of this interest. Astronomy studies the origin, development, and composition of the stars and other celestial objects, and the laws of their movement in the universe. Over time, this science would help calm the fears of the unknown that disturbed the earliest peoples.

The universe is made up of billions of stars, planets, and other cosmic elements that are constantly expanding.

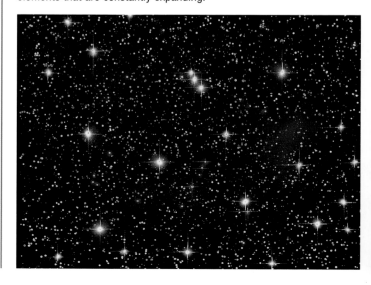

The stars and planets are arranged in galaxies. These galaxies are so large that light from them may take thousands or even millions of years to travel from one end of a galaxy to the other.

Most people believed that the movements of the planets had an influence over the Earth and over humans. Many thought that each person's destiny was mapped out in the sky. From this desire to "read" the messages of the night sky came a new science that paralleled astronomy. This science, known as astrology (the reading of stars and planets), was of great interest to people in the ancient world.

GEOCENTRISM AND HELIOCENTRISM

The Greek civilization developed classical astronomy more than any other ancient people. The Greeks looked at the universe in terms of geometry rather than the supernatural. Both the Greeks and other ancient peoples saw the cosmos as a geocentric universe—that is, they believed the Earth was located at the center of the universe and that the other celestial bodies rotated around it. This idea, made famous by Ptolemy (A.D. 100–170), lasted until the 16th century. Religious and political groups worked to keep this belief (geocentrism) in place for many years.

After a journey of seven months, the NASA space probe *Spirit* landed on Mars on January 4, 2004. The perilous descent through the Martian atmosphere took six minutes. The probe's main goals were to explore the "red planet," while looking for water and traces of life; to study the geology of Mars; and to take high-quality pictures of the planet's surface.

For this reason, Western astronomy did not progress during the Middle Ages (from the fall of Rome in 375 to the mid-15th century). During this time, only Hindu and Arab astronomers kept science alive. The Arabs compiled new catalogs of stars in the 9th and 10th centuries and developed charts that showed how the planets moved. The archives of Arabian astronomers and Arab works on astronomy later helped Western astronomy grow when they were translated into Greek and Latin.

The theory of geocentrism remained a belief of many until the 16th century. At this time, the history of astronomy took a drastic turn thanks to Nicolaus Copernicus (1473–1543), the Polish astronomer who introduced the heliocentric theory, which suggested that the sun was at the center of the universe and the rest of the planets, including the Earth, moved around it.

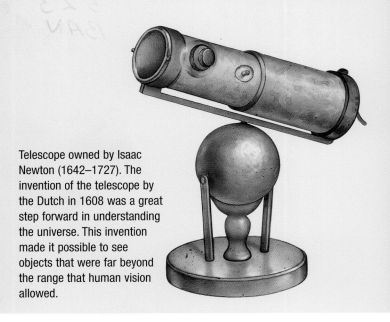

Telescope owned by Isaac Newton (1642–1727). The invention of the telescope by the Dutch in 1608 was a great step forward in understanding the universe. This invention made it possible to see objects that were far beyond the range that human vision allowed.

THE RESURGENCE OF ASTRONOMY

Until the beginning of the 17th century, astronomers had looked at the sky with the naked eye alone. In Holland, technology made a major contribution to astronomy with the invention of the telescope, attributed to Hans Lippershey (1570–1619). But it wasn't until 1609 that the Italian astronomer Galileo Galilei (1564–1642) used it to study the heavens. For the first time, someone was viewing the night sky with something that surpassed the ability of the human eye.

Thanks to his studies of the planets with his telescope, Galileo found evidence to support the heliocentric theory of Copernicus, which had been largely ignored until then. Galileo's efforts to make his system known brought him before a religious court. Although he was forced to deny his beliefs and his writings, people's interest in his theory could not be suppressed.

Once the heliocentric theory was accepted, the German astronomer Johannes Kepler (1571–1630) confirmed in 1609 that Mars did not rotate around the sun in a circle or a combination of circles as all Western astronomers, including

Copernicus, had believed. Mars moved around the sun in an elliptical (oval-shaped) orbit. Kepler then showed that all planets followed elliptical orbit patterns.

The findings of English scientist Isaac Newton (1642–1727) marked the end of the transition from ancient to modern astronomy. Newton came up with a simple principle to explain Kepler's laws of planetary motion—the force of attraction between the sun and the planets. This force depends on the masses of the sun and the planets and on the distances between them. Newton's mathematical discovery is called the "law of universal gravitation."

The advance of technology has allowed astronomy to develop in spectacular ways in recent years. Today, we look at the universe not only by visual means, but we can also analyze radiation that comes from space.

"One small step for man, one giant leap for mankind." These words were said in 1969 by Neil Armstrong, who was the first man to set foot on the moon, accompanied by Edwin "Buzz" Aldrin.

THE GREAT ADVANCES OF RECENT CENTURIES

In the 19th century, a new instrument called the spectroscope brought new information about the chemical makeup of celestial bodies and new data about their movements. During the 20th century, astronomical observation based on the analysis of visible light gave way to other ways of viewing the universe and celestial bodies. Also during this period, there were important increases in the size and power of telescopes.

Meteorites may have caused some of the catastrophic events of the past, such as the extinction of dinosaurs at the end of the Cretaceous Period. It is believed that this massive die-off was started by the crash of a meteorite more than 6 miles (9.7 km) in diameter 65 million years ago.

In 1905, Albert Einstein published his special theory of relativity. This theory is the key to understanding the way the universe was formed. It is also important for understanding how space and time relate to each other. Eleven years later, Einstein published an expanded version of his work in a book about the general theory of relativity.

Technological advances came from rapid leaps in engineering and technical know-how in the second half of the 20th century (such as faster computers, more sophisticated orbiting telescopes, and space probes sent to explore planets and other bodies). These new technologies have allowed astronomy to go through a revolution. These technological breakthroughs have given us reasons to hope that some of the mysteries that still remain in the universe will eventually be uncovered.

AND ENERGY WAS CONVERTED TO MATTER

In 1929, American astronomer Edwin Hubble was able to show that the galaxies are moving away from each other. This observation made it clear that the universe is expanding. If this is the case, then it would seem logical to suppose that, if we were to go back in time, galaxies would be even closer to each other, and the universe would be even more dense until all matter would be joined together. This matter would be unstable, and would cause an initial explosion known as the Big Bang.

THE FUTURE OF THE UNIVERSE

Since the Big Bang, the universe has continued to expand. Scientists believe that the future holds two possibilities: the universe will keep expanding indefinitely (the theory of a perpetual universe), or that a time will come when the expansion relaxes, and the universe will again contract (the theory of an oscillating or pulsing universe).

1 time begins
a great explosion, or "Big Bang," occurs. At this moment, the density of the universe is infinite, and its volume is zero. In other words, the universe is infinitely small in size, and its temperature is infinitely high.

2 rapid expansion
in a small fraction of a second, the cosmos rapidly expands from the size of an atom to the size of a grape.

10^{-43} seconds	10^{-32} seconds		10^{-6} seconds	1 second	3 minutes
temperature	10^{27} °C		10^{13} °C		10^{8} °C

1   2   3   4   5   6

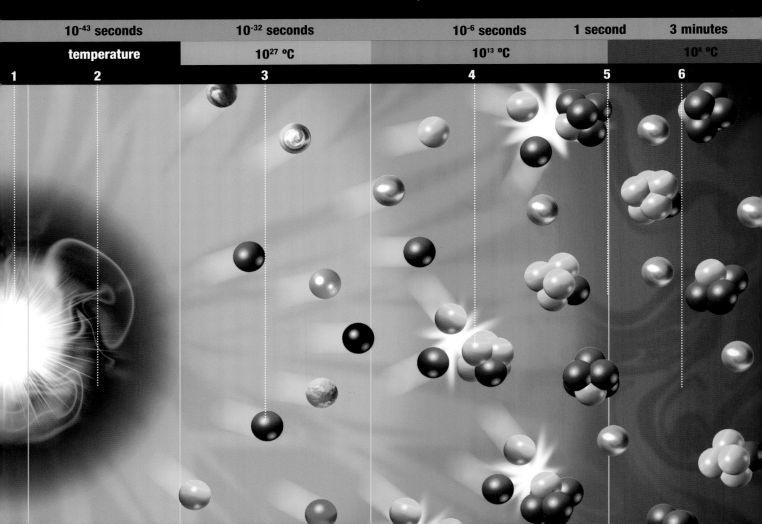

3 the first elementary particles appear

the universe is a boiling soup of elementary particles (electrons, quarks, and other particles). A big expansion of the universe occurs.

4 the first atomic nuclei appear

a rapid cooling of the cosmos allows quarks and other elementary particles to group into atomic nuclei, the first of the light elements (hydrogen and helium), and then, over time, into the heavier elements.

5 the universe is one second old

6 the universe is an extremely hot nebula

the temperature is still too hot for atoms to form.

7 matter finally forms

electrons combine with protons and neutrons to form atoms (matter), made up mainly of hydrogen and helium. Light shines in the universe for the first time.

8 the first galaxies form

gravity makes hydrogen and helium gas collapse and form giant clouds that become the first galaxies. Small clusters of gas collapse to form the first stars.

9 the forces of gravitational attraction come into play

gas and dust break into a large number of galaxies. These galaxies turn around each other and, because they are unstable, break up into billions of stars.

10 the present day

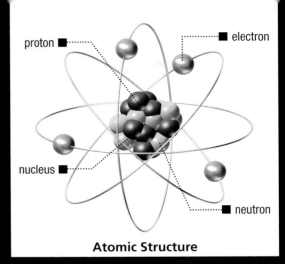

Atomic Structure

proton — electron — nucleus — neutron

The atom is the basic unit of matter. It is made up of a nucleus, which consists of protons (particles with mass and a positive electric charge), neutrons (which have mass, but no electric charge), and an outer shell, where electrons (negatively charged particles that have no mass) are found.

300,000 years	one billion years	15 billion years
10.000 ºC	-200 ºC -270 ºC	
7	8 9	10

IN AN ARM OF THE GALAXY

Galaxies are huge groupings of stars, planets, gases, and dust that are held together by a force called gravity. Scientists believe that the known universe contains a billion galaxies, and that each one of these may be made up of billions of stars and other star-like objects. Our solar system is part of a galaxy known as the Milky Way, which is made up of more than 100 billion stars.

1 Spiral arms
form because stars move at a different speed from the center of the Milky Way galaxy.

2 Galactic center
cannot be seen because, between this zone and the Earth, there are dark clouds made up of large masses of dust that block our direct line of sight to the center of the galaxy.

3 Solar system
is a group of celestial bodies that are held together by the sun.

4 Nebulae
are huge clouds made of light gases (hydrogen and helium) and cosmic dust (carbon, iron, silicon).

5 Arm of Carina

6 Arm of Sagittarius

7 Arm of Orion
here is where the sun and the stars we see are located.

8 Arm of Perseus

SOME FACTS ABOUT THE MILKY WAY

Traveling at the speed of light, it would take about 100,000 years to cross the Milky Way from one end to the other. The orbit of the sun around the center of the galaxy has a diameter of 60,000 light years, and the sun takes 240 million light years to make one orbit around the galactic center. The distance between the sun and the center of the galaxy is 25,000 light years.

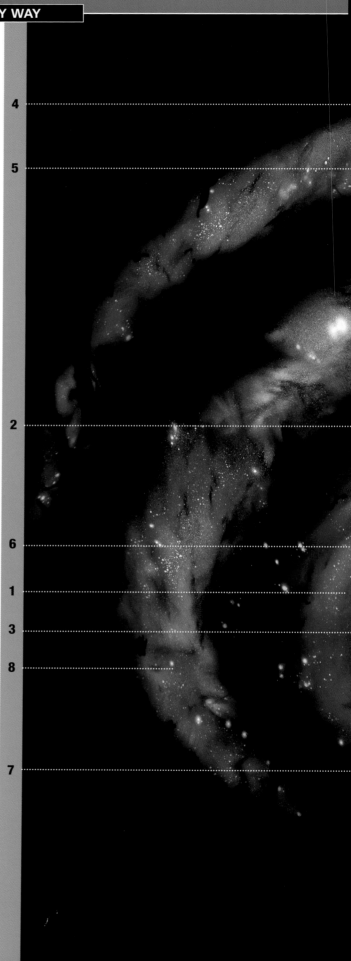

Supernova

A supernova is the explosion of a star.
During this event, an exploding star may
shine 100 billion times more brightly
than the sun for a few days
before it vanishes forever.

EARTH'S SOLAR SYSTEM

Our solar system consists of a medium-sized star that we call the sun, and the planets, including the Earth, that orbit around it. It also includes more than 60 satellites of those planets, as well as many comets, asteroids, meteorites, and space between the planets. In relation to the sun, the first four planets (Mercury, Venus, Earth, and Mars) are called the "inner planets" because they are closest to the sun, while Jupiter, Saturn, Uranus, Neptune, and Pluto are called the "outer planets."

1 Mercury

during the day, temperatures can reach 800°F (427°C), and at night, they get as low as -300°F (-184°C). Mercury is the smallest of the inner planets, and has no satellites.

2 Venus

is the brightest celestial object after the moon. Venus spins in a direction opposite to that of the other planets in the solar system: On a day in Venus, the sun rises in the west and sets in the east.

3 Earth

is the largest of the "inner planets," and the only one on which life and liquid water are known to exist.

4 Mars

the "red planet" has the highest volcanic formations in the solar system: Mount Olympus is 82,000 feet (24,994 meters) high.

The rings of Saturn

Saturn's complex series of rings is formed from objects of different sizes, including frozen ice particles (particularly at the outer edges) and ice-covered rock up to 3,200 feet (975 meters) in diameter (especially in the inner sections).

5 Jupiter
is the largest planet in the solar system, has the greatest mass, and rotates at the highest speed.

6 Saturn
has a belt formed by rings made up of dust and fine particles that orbit the planet as if they were small satellites.

THE ASTEROID BELT
Asteroids are rocky objects that orbit around the sun. Most of them are found in the Asteroid Belt, which lies between Mars and Jupiter. Asteroids are also called "small planets." They range in size from small particles to over 620 miles (998 km) in diameter. It is believed that they are made of material that was never able to join together to form a new planet.

7 Uranus
is unusual because it leans to one side. This strange position may be the result of a collision with another planetary body during the early history of the solar system.

8 Neptune
has the strongest winds in the solar system; wind speeds reach more than 1,200 miles (1,931 km) per hour.

9 Pluto
is actually inside the orbit of Neptune during part of its path around the sun. The time it takes to move around the sun is the longest of all the planets.

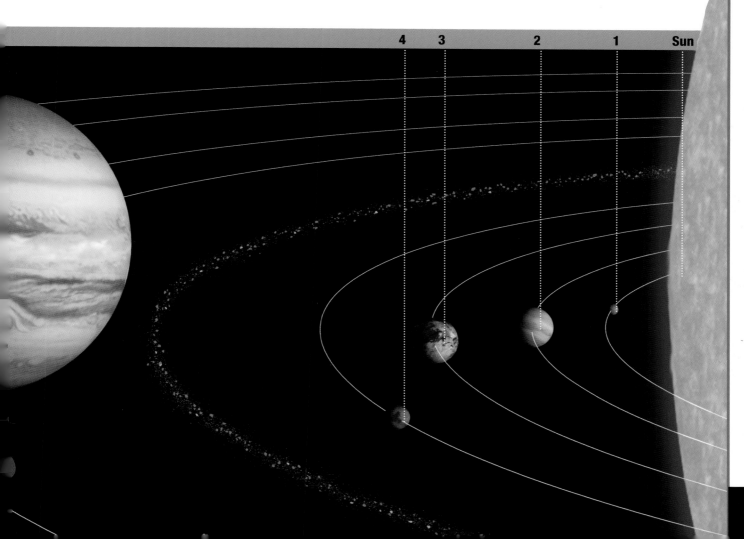

A FANTASTIC BALL OF FIRE

The sun is a star located at the heart of our solar system. Its huge gravitational force keeps the planets, asteroids, and comets in their orbits. The interior of the sun is a nuclear reactor that radiates light and heat through the whole solar system. The sun's surface area is made up of about 73% hydrogen, 25% helium, and 2% other elements. Its mass is 333,000 times greater than that of Earth.

1 Core
is the source of the sun's energy, which is generated as hydrogen atoms fuse to become helium nuclei.

2 Radioactive zone
energy is transmitted in this zone by electromagnetic waves that interact continually with matter. Its temperature is lower than that of the core.

3 Convective zone
in this zone, currents move up and down as they carry heat emissions to the photosphere; the temperature is 2,700,000°F (1,499,982°C).

4 Photosphere
is the layer from which just about all visible light is emitted from the sun. For this reason, it represents what normally is called the "surface" of the sun; its temperature is around 10,500°F (5,816°C).

5 Chromosphere
is an outer reddish layer, made of hydrogen, that rises 620 miles (998 km) over the photosphere; its temperature is between 7,000 and 14,500°F (between 3,871 and 8,038°C).

6 Corona
is the outermost layer of the sun. It can reach a height of over 2 million miles (3.2 million km), at which point it gradually turns into solar wind. It may reach a temperature of up to 3.6 million°F (2 million°C).

7 Sunspots
arise where the sun's magnetic field is concentrated, so that these areas are cooler (and appear darker) than the surrounding areas.

8 Protuberances
are powerful arched jets of hot gas that form when magnetic energy stored in the chromosphere is suddenly released. They can reach up to 12,428 miles in height.

9 Filaments
when they are seen with the solar disk as a background, protuberances look like dark bands called filaments.

10 Spicules
are vertical jets of gas, capable of rising to a height of some 6,200 miles (9,978 km) above the surface. There are about 100,000 spicules on the sun's surface at any given moment, but they only last for a few minutes.

11 Granulations
are small, brilliant spots that have an average life of a few minutes, and can reach up to 930 miles (1,497 km) in diameter. These are due to the existence of columns of very hot gas that rise up from the photosphere.

Solar flares and polar auroras

The chromosphere of the sun launches invisible jets of ionized gases, called "solar wind." When these gases arrive at the Earth's atmosphere, they move rapidly from one pole to the other, creating the polar auroras (aurora borealis in the north and aurora australis in the south) and electrical storms, causing interference with radio waves all over the planet.

THE DEATH OF THE SUN

The sun has been active for 4.6 billion years, and has enough fuel to last another 5 billion years. At the end of its life, the sun will begin to mix helium with heavier elements and will begin to swell. At the final stage, it will be so large that it will absorb Earth and, 1 billion years later, will collapse. That will be the end of this star.

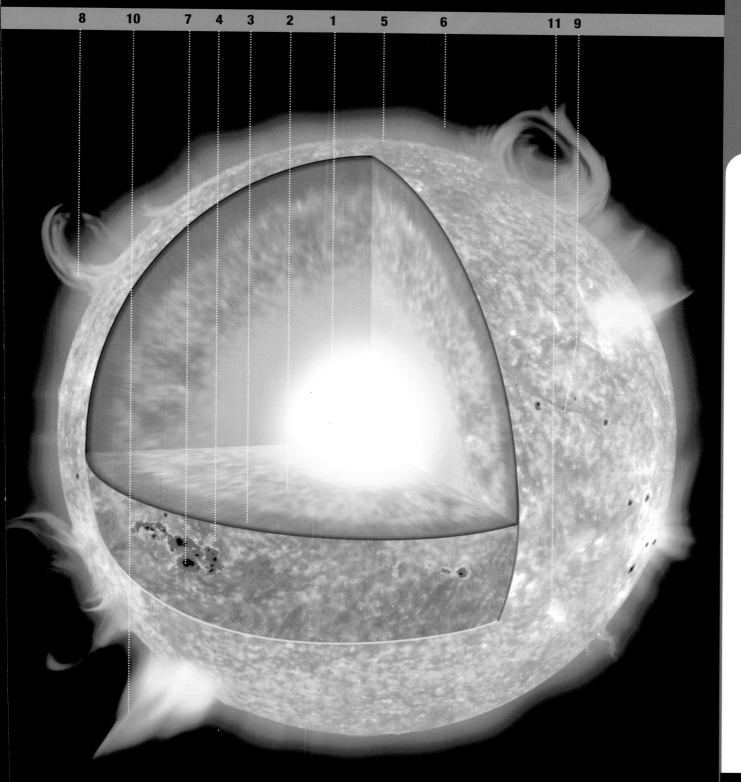

8   10   7   4   3   2   1   5   6   11   9

WANDERERS OF THE SOLAR SYSTEM

Comets are small celestial bodies that orbit the sun. When their orbits carry them too close to the sun, their surface ice evaporates, forming a long, glowing tail. The core of a comet, which contains nearly all of its mass, is made of ice, organic compounds (carbon, methane, ammonium, and other molecules), gases (hydrogen, oxygen, nitrogen), and dust. It is essentially a kind of "dirty snowball" in space. A comet's tail gets longer as it nears the sun, and it may reach millions of miles in length.

HALLEY'S COMET

In 1682, the astronomer Edmond Halley found that the comets seen in 1531, 1607, and 1682 followed the same path, and he concluded that they were all the same comet, which would return in about 1758. His prediction was correct, but he did not live to see it come true, since he died in 1742.

Orbit of a comet ■
may be open (the comet circles the sun and recedes, never to return again) or closed (the comet will return).

Sun ■
its huge gravitational attraction holds the comets in their orbits.

Direction of the tail ■
because of solar wind, the tail is always extended away from the sun, even when the comet is moving away from the central star.

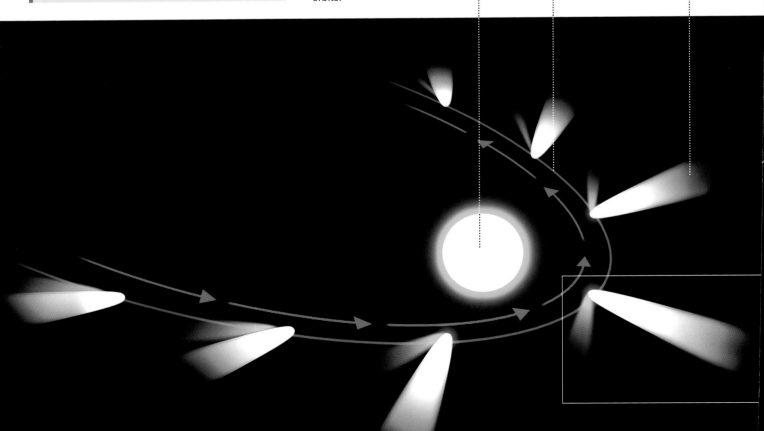

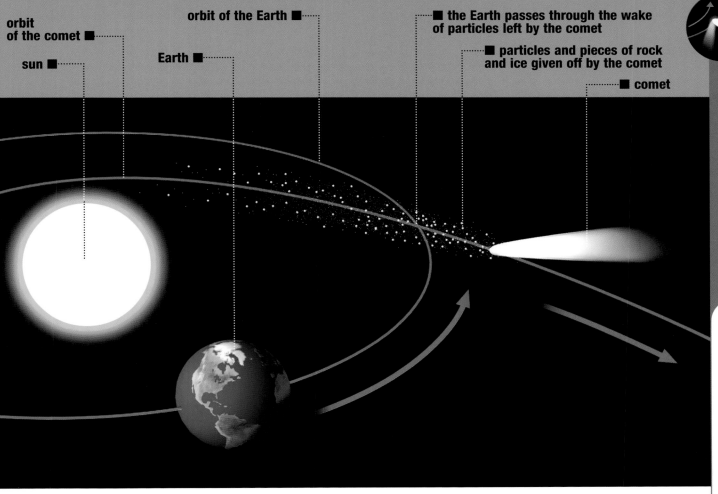

orbit of the comet ■
sun ■
orbit of the Earth ■
Earth ■
■ the Earth passes through the wake of particles left by the comet
■ particles and pieces of rock and ice given off by the comet
■ comet

In its path, a comet leaves large amounts of small pieces of material. When the Earth crosses a comet's orbit, these fragments fall into the atmosphere in the form of shooting stars.

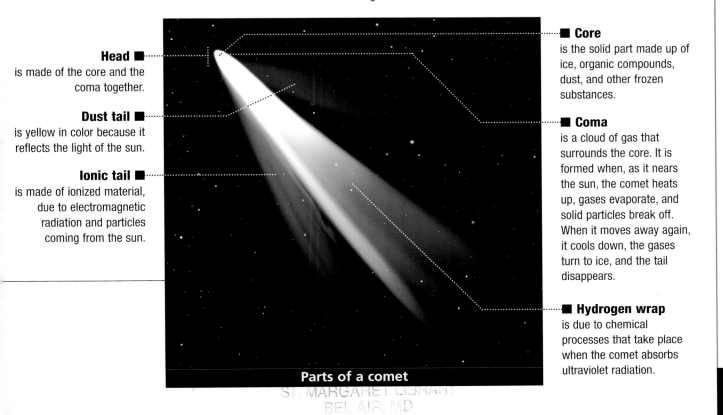

Head ■
is made of the core and the coma together.

Dust tail ■
is yellow in color because it reflects the light of the sun.

Ionic tail ■
is made of ionized material, due to electromagnetic radiation and particles coming from the sun.

■ **Core**
is the solid part made up of ice, organic compounds, dust, and other frozen substances.

■ **Coma**
is a cloud of gas that surrounds the core. It is formed when, as it nears the sun, the comet heats up, gases evaporate, and solid particles break off. When it moves away again, it cools down, the gases turn to ice, and the tail disappears.

■ **Hydrogen wrap**
is due to chemical processes that take place when the comet absorbs ultraviolet radiation.

Parts of a comet

A LIVING PLANET

The Earth is one of the nine planets in our solar system. It is the third closest planet to the sun and the largest of the interior planets. Its position and its mass make it a privileged planet, with an average temperature of 60°F (16°C), water in liquid form, and an oxygen-rich atmosphere— all important conditions for the development of life. The Earth's interior is divided into several layers that have different chemical and seismic properties.

Atmosphere ■ protects us from external agents such as ultraviolet radiation and meteorites, and contains the oxygen that we breathe.

Emerged surface of the crust ■ is where we find geographical features (such as mountains, valleys, plains, rivers, lakes, and glaciers).

Hydrosphere ■ is the layer of water that, in the form of oceans, seas, lakes, rivers, and ice, covers 71% of the Earth's surface.

Lithosphere ■ includes the crust and the top part of the mantle. Appears to be divided into tectonic plates that move slowly over the asthenosphere.

Asthenosphere ■ is a layer of the upper mantle whose rocks are in a partly fluid state. The tectonic plates "float" over this area. This is the main source of magma.

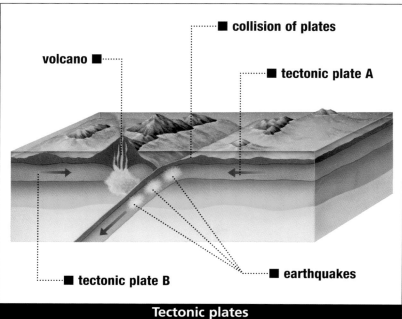

collision of plates

volcano ■

tectonic plate A

tectonic plate B

earthquakes

Tectonic plates

Tectonic plates move with respect to each other; this movement produces great friction where their edges meet. This friction releases a large amount of energy in the form of earthquakes and volcanoes.

THE EARTH IS RENEWED

The Earth is approximately 4.6 billion years old, but its surface is relatively young (500 million years). Erosion and tectonic processes destroy and renew the surface of the Earth, and thereby rid the planet of nearly all traces of the earliest surface geology.

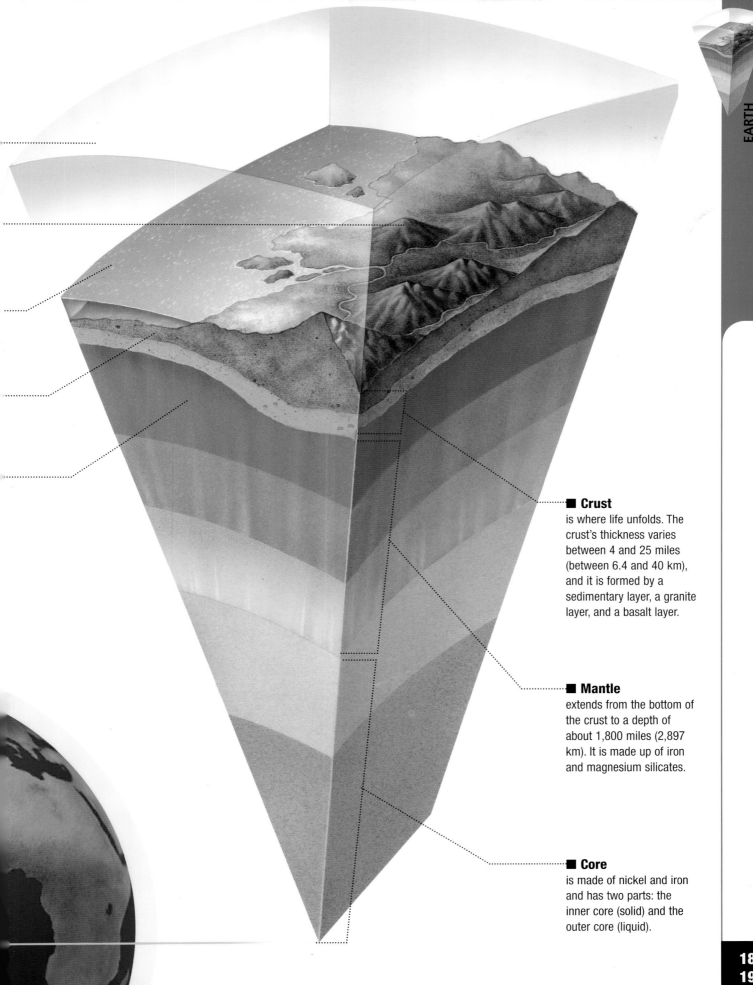

■ **Crust**
is where life unfolds. The crust's thickness varies between 4 and 25 miles (between 6.4 and 40 km), and it is formed by a sedimentary layer, a granite layer, and a basalt layer.

■ **Mantle**
extends from the bottom of the crust to a depth of about 1,800 miles (2,897 km). It is made up of iron and magnesium silicates.

■ **Core**
is made of nickel and iron and has two parts: the inner core (solid) and the outer core (liquid).

THE INSEPARABLE COMPANION OF THE EARTH

The moon is Earth's only satellite. The average distance between these two bodies is approximately 236,000 miles (379,805 km), and the moon's orbit around Earth lasts about 28 days. It has no water or air, and so cannot support life. It is believed that the moon was created by a collision between the newly formed Earth with an object roughly the size of Mars. The impact would have torn away enough material from Earth to form the moon, which remained in orbit around the Earth.

THE HIDDEN FACE OF THE MOON

Since the movement of the moon around the Earth and its rotation take about the same amount of time, the moon always shows the same part of its surface when viewed from the Earth. The other side of the moon is always invisible to an observer standing at any place on Earth.

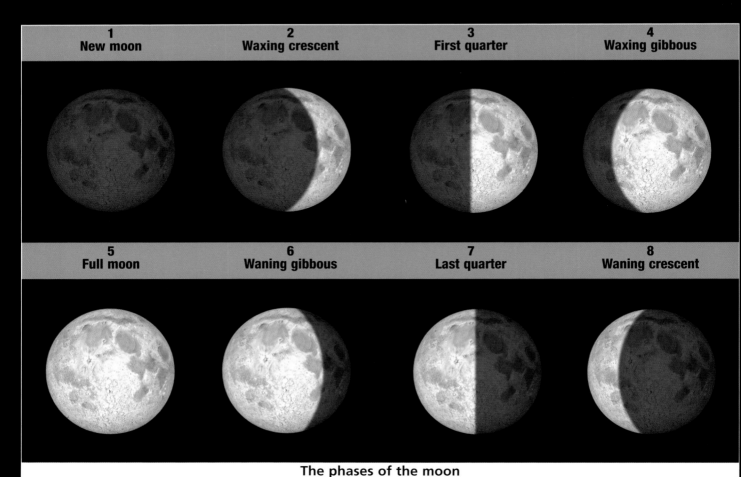

1 New moon	2 Waxing crescent	3 First quarter	4 Waxing gibbous
5 Full moon	6 Waning gibbous	7 Last quarter	8 Waning crescent

The phases of the moon

The moon does not have its own light. It shines because it reflects the light of the sun. Depending on where the moon is in its orbit around Earth, the part that lights up seems to take on a shape, which we call a "phase."

1 Visible face of the moon
is the part of the moon that we always see from Earth.

2 Seas
are the dark areas; these are formed by solidified lava.

3 Craters
formed billions of years ago by meteor impacts.

4 Moon landings by the United States

5 First moon landing
is the place where a human being first set foot on the moon.

6 Moon landings by the former Soviet Union

7 Sea of Cold

8 Sea of Rains

9 Sea of Serenity

10 Sea of Tranquillity

11 Sea of Humors

12 Sea of Clouds

13 Copernicus Crater

THE MOON

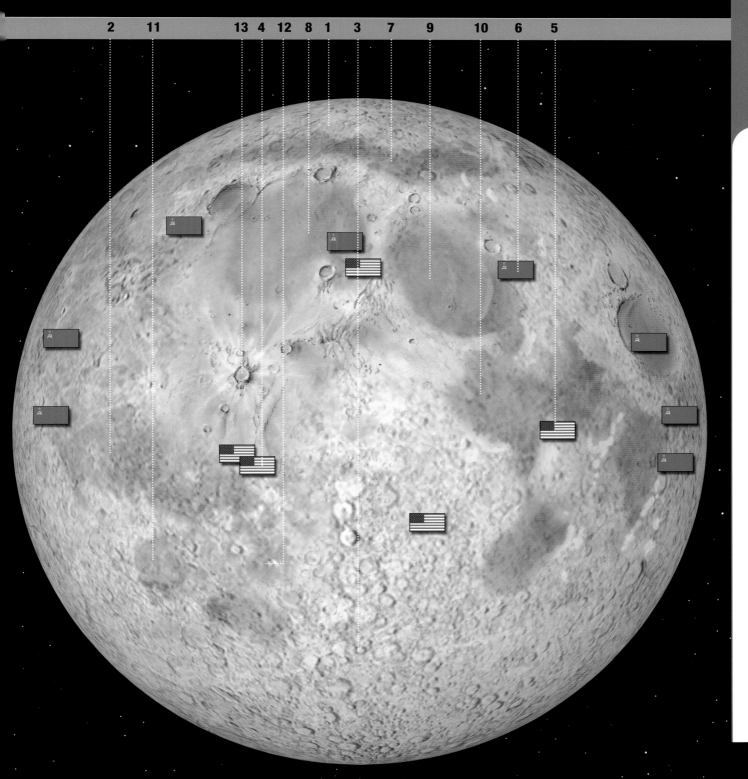

PLAYING HIDE AND SEEK

Lunar and solar eclipses occur because of the way the moon revolves around the Earth. When the Earth, the moon, and the sun are all in a line, the shadow of the Earth sometimes falls on the moon and produces a lunar eclipse. At other times, the shadow of the moon falls on the Earth, causing a solar eclipse. When the cone of the shadow does not completely cover the moon or the Earth, this is called a partial eclipse. A full eclipse occurs when the shadow covers them completely.

WHY ARE ECLIPSES RARE?

During most of the times when the moon is between the sun and the Earth (during the new moon phase), or when the Earth is located between the moon and the sun (during the full moon phase), the moon is on top of or beneath the Earth's orbital plane. This keeps the three bodies from getting into the perfect line needed to produce a lunar or solar eclipse.

1 Umbra
is the area of the moon that is completely darkened by the shadow of the Earth.

2 Penumbra
is the area of the moon that is partially darkened by the shadow of the Earth.

3 Sun

4 Earth
is positioned between the sun and the moon, causing a shadowy area in space.

5 Sunlight

6 Moon
is darkened totally or partially when it enters the area of the Earth's shadow.

7 Orbit of the moon

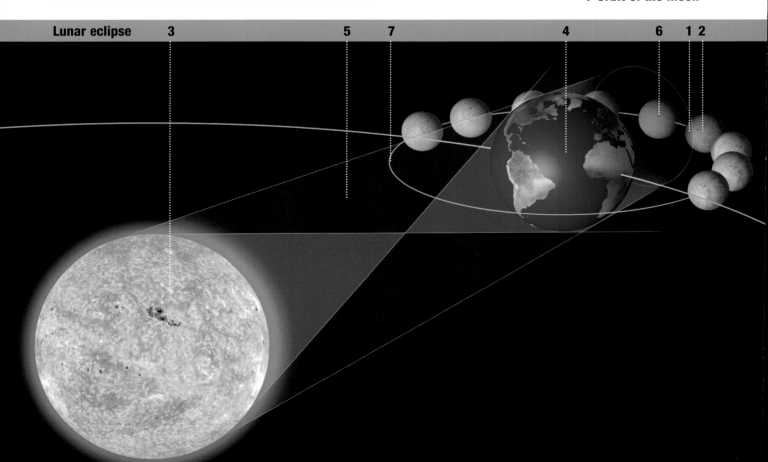

Lunar eclipse　　3　　　　　5　7　　　　　　4　　　　6　1 2

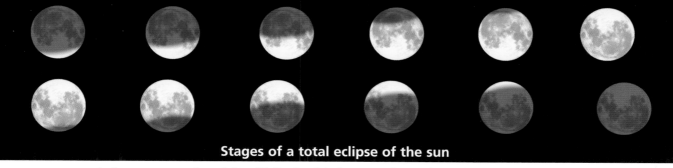

Stages of a total eclipse of the sun

As the moon moves through its orbit, it enters into the cones of penumbra and umbra, then emerges, first from the umbra, and then from the penumbra.

Stages of a total eclipse of the moon

During a total eclipse, the moon lines up exactly with the center of the sun, producing a corona effect. At this moment, with a telescope, we can observe solar flares on the edges of the moon.

1 Umbra
is the area of the Earth that is completely darkened by the moon's shadow.

2 Penumbra
is the area of the Earth that is partially darkened by the moon's shadow.

3 Sun

4 Earth
upon entering the shadow area of the moon, the sun is totally or partially covered.

5 Moon
inserts itself in broad daylight between the sun and the Earth, causing an area of shadow.

6 Sunlight

7 Orbit of the Earth

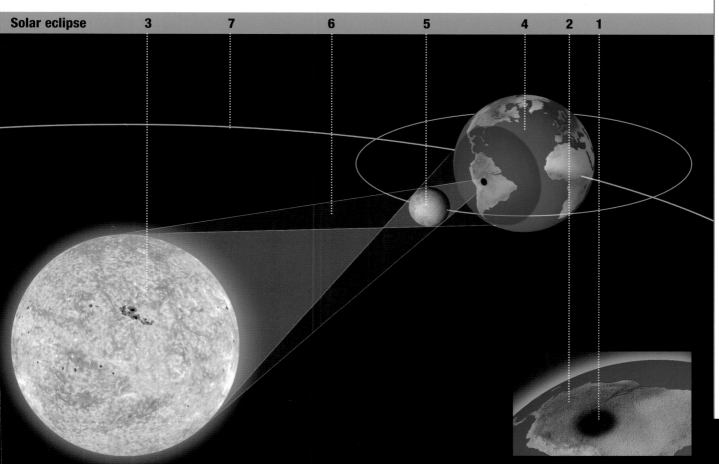

Solar eclipse   3   7   6   5   4   2   1

STARS OF LEGEND

Constellations are groups of stars in recognizable shapes, which were invented by humans who wanted to find their ancestors, gods, legends, animals, and mythological beings in them. The constellations do not reflect real groups of stars. In fact, the stars that make up a constellation may actually be hundreds of light years away from each other, even though they may look like they are close together. Today, 88 constellations are identified; many of them are visible from both hemispheres of the Earth, and others can be seen from only one hemisphere.

CONSTELLATIONS OF THE NORTHERN HEMISPHERE

Pegasus ■

Cygnus ■

Lyre ■

Perseus ■

Aries ■

Andromeda ■

Pisces ■

Hercules ■

Virgo ■

Leo ■

■ Big Dipp●

■ Little Dipper

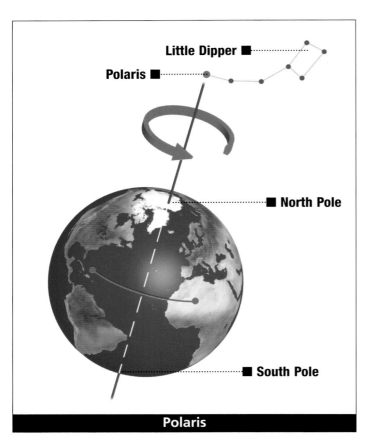

Little Dipper ■

Polaris ■

■ North Pole

■ South Pole

Polaris

This is the star that is found at the spot where the Earth's axis in the Northern Hemisphere extends out into space. It forms part of the Little Dipper. Before the invention of the compass, Polaris was the only reference point to indicate which way was north.

THE CONSTELLATIONS OF THE ZODIAC

There are 12 famous constellations of the Zodiac. They are arranged throughout the path that the sun travels during the year. Due to this movement, the sun covers a different constellation every month, which means that this particular constellation cannot be viewed, because the area of the sky closest to the sun is too bright for stars to be seen.

CONSTELLATIONS OF THE SOUTHERN HEMISPHERE

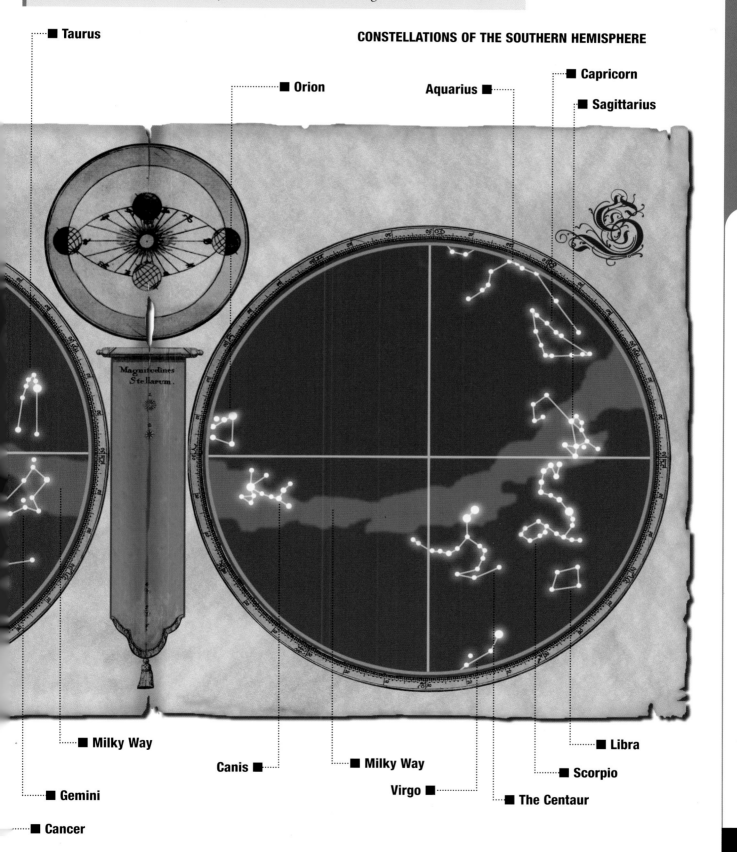

■ Taurus

■ Orion

Aquarius ■

■ Capricorn

■ Sagittarius

Magnitudines Stellarum.

■ Milky Way

Canis ■

■ Milky Way

Virgo ■

■ Libra

■ Scorpio

■ The Centaur

■ Gemini

■ Cancer

HUBBLE: AN EYE IN SPACE

The Hubble Space Telescope was placed in orbit on April 25, 1990. It is an artificial satellite that circles the Earth every 90 minutes, at an altitude of 375 miles (604 km). The biggest difference compared with other satellites is that Hubble directs its instruments toward space rather than Earth. The main structure of Hubble is not very different from that of an Earth-based telescope: It is a tube, inside of which is a large mirror more than 8 feet (2.4 meters) in diameter.

■ Door
protects the mirrors from contamination and cosmic dust.

Secondary mirror ■
reflects light toward the instruments located behind the primary mirror.

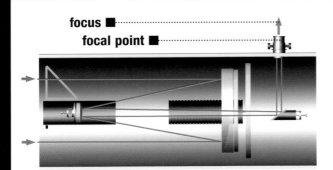

focus ■
focal point ■

The telescope

The telescope was invented in Holland in the year 1608, but its invention was regarded as a military secret. Galileo Galilei heard about this invention and decided to design and build his own telescope. Thanks to the telescope, he made great discoveries in astronomy, including his observation of four of the moons of Jupiter on January 7, 1610.

WHY A SPACE TELESCOPE?

Earth's atmosphere, which is transparent to visible light, absorbs most of the "electromagnetic radiations": part of the infrared (IR) radiation, all of the ultraviolet (UV) radiation, X rays, and gamma rays. These forms of light are useless to telescopes on Earth. Even visible light is not entirely useful since changes in the atmosphere cloud telescope images if they are highly magnified. The Hubble telescope, in low Earth orbit, gets rid of these problems inside the visible spectrum and for part of the UV range.

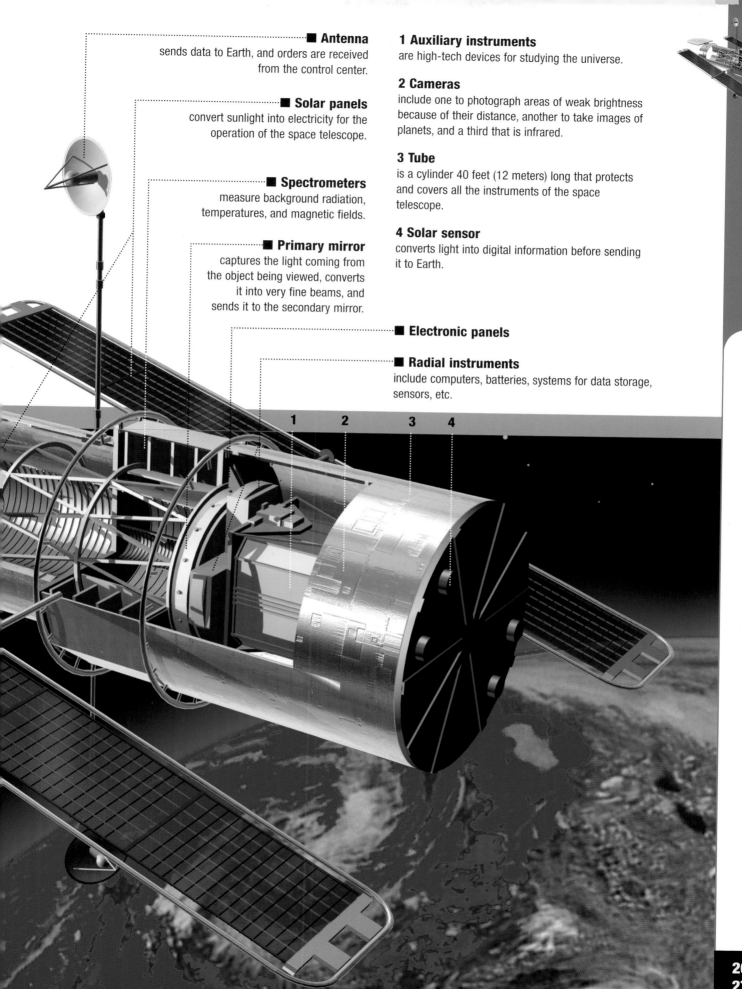

■ Antenna
sends data to Earth, and orders are received from the control center.

■ Solar panels
convert sunlight into electricity for the operation of the space telescope.

■ Spectrometers
measure background radiation, temperatures, and magnetic fields.

■ Primary mirror
captures the light coming from the object being viewed, converts it into very fine beams, and sends it to the secondary mirror.

1 Auxiliary instruments
are high-tech devices for studying the universe.

2 Cameras
include one to photograph areas of weak brightness because of their distance, another to take images of planets, and a third that is infrared.

3 Tube
is a cylinder 40 feet (12 meters) long that protects and covers all the instruments of the space telescope.

4 Solar sensor
converts light into digital information before sending it to Earth.

■ Electronic panels

■ Radial instruments
include computers, batteries, systems for data storage, sensors, etc.

THE CROWN JEWEL OF SPACE TRAVEL

NASA's space shuttle is the first reusable spaceship. First proposed in May 1972 as an economical way to put satellites into orbit and bring them back to Earth, the first space shuttle launch did not happen until April 1981. Each shuttle has a projected useful life of 100 launches. Of the six shuttles placed into operation, the *Challenger* and the *Columbia* were destroyed accidentally in flight. All of the members of each crew were killed as a result of the tragedies. *Discovery*, *Endeavor*, and *Atlantis* are still in service, and the *Enterprise* hasn't been used yet.

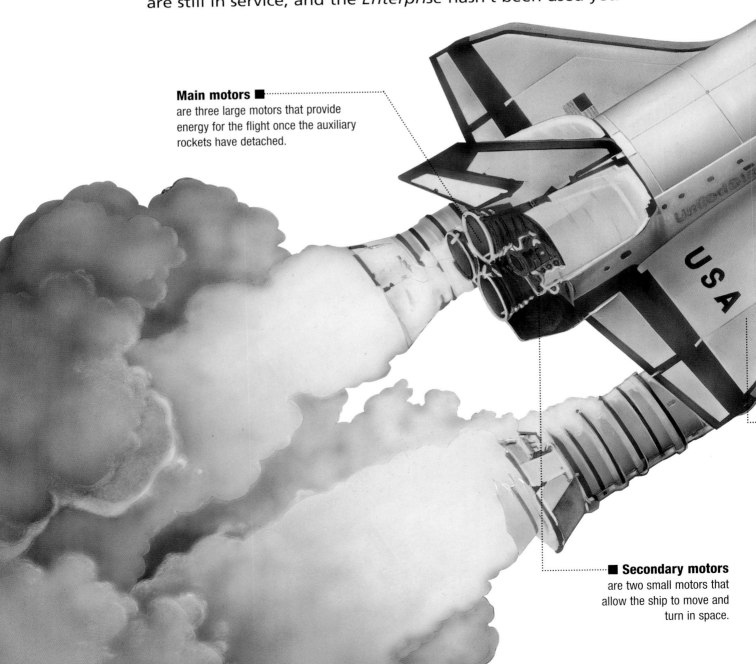

Main motors ■
are three large motors that provide energy for the flight once the auxiliary rockets have detached.

■ **Secondary motors**
are two small motors that allow the ship to move and turn in space.

Cabin ■
is where the pilot and copilot sit; there are also seats for the rest of the crew during takeoff and reentry into the atmosphere.

External fuel depot ■
holds the fuel needed for the first part of the launch into space. The ship is detached from it at an altitude of 56 miles (90 km).

Compartments ■
are where the crew members spend their time and where the everyday activities of the space mission take place.

■ **Auxiliary rocket propellants**
make the ship rise to an altitude of 25 miles (40 km). At this height, they are detached and fall with parachutes into the sea, where they are recovered.

■ **Access hatch to the crew area**

■ **Security hatch**
is used only in case of emergency.

■ **Cargo hold**
has room to carry nearly 30 tons of useful equipment.

■ **Cargo hold deck**
has lockgates that open for loading and unloading from the cargo hold.

■ **Thermal protection tiles**
are made of ceramic and absorb the heat produced upon reentry of the ship into the atmosphere; keeps the interior temperature within limits to protect the crew members.

THE SPACE SHUTTLE'S FUEL

The fuel that the space shuttle uses is a mixture of liquid hydrogen and oxygen. The main motors take fuel from the main tank at a rate of 47,000 gallons (178,000 liters) of hydrogen per minute and 17,000 gallons (64,000 liters) of oxygen per minute. A car could make 60 trips around the world with this amount of fuel.

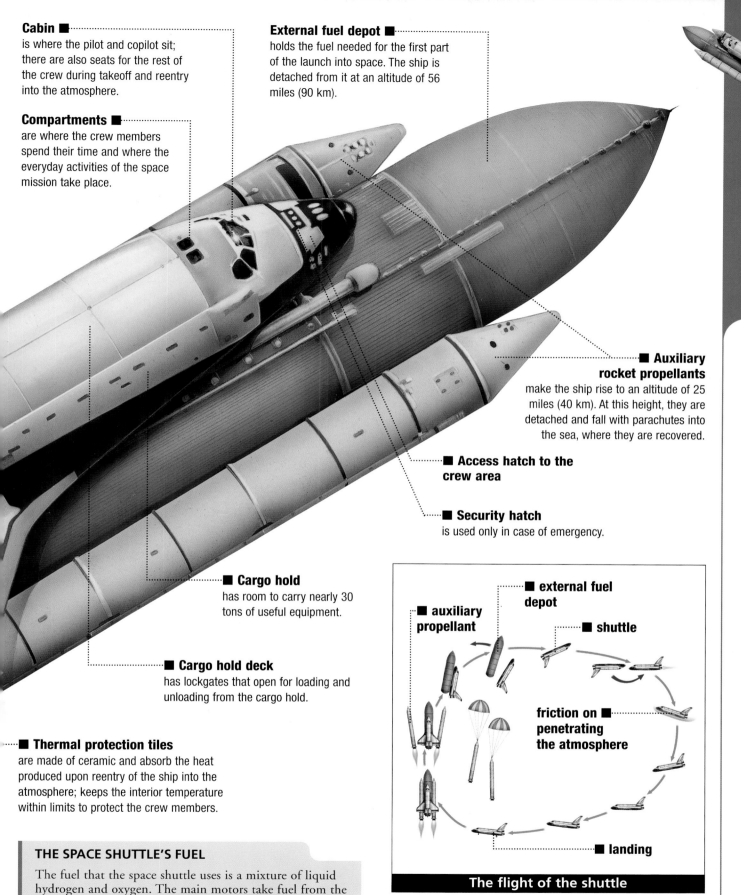

■ **auxiliary propellant**

■ **external fuel depot**

■ **shuttle**

friction on ■ **penetrating the atmosphere**

■ **landing**

The flight of the shuttle

The space shuttle takes off like a rocket (helped by auxiliary propellants that are recovered later and by a fuel depot that is lost) and lands like an airplane.

DID YOU KNOW?

THE PLANETS OF THE SOLAR SYSTEM

INNER PLANETS	Mercury	Venus	Earth	Mars
Distance to the sun (millions of miles)	36.0	67.2	92.96	141.6
Diameter (miles)	3,031	7,520	7,926	4,217
Period of rotation around the sun (days)	87.97	224.70	365.26	686.98
Velocity of orbit around the sun (miles/sec)	29.76	21.77	18.51	14.99
Period of rotation	58 days, 16 hours	243 days, 14 min	23 days, 56 min	24 days, 37 min
Mass (Earth = 1)	0.055	0.81	1	0.11
Density (water = 1)	5.43	5.25	5.52	3.95
Temperature (°F on the surface)	-292°/806°	869°	-94°/131°	-184°/77°
Number of moons	-	-	1	2

OUTER PLANETS	Jupiter	Saturn	Uranus	Neptune	Pluto
Distance to the sun (millions of miles)	483.6	886.7	1,784	2,794	3,675
Diameter (miles)	88,846	74,900	31,763	30,775	1,419
Period of rotation around the sun (days)	11.86	29.46	84.01	164.79	248.54
Velocity of orbit around the sun (miles/sec)	8.12	5.99	4.23	3.37	2.95
Period of rotation	9 hours, 55 min	10 hours, 40 min	17 hours, 14 min	16 hours, 7 min	6 days, 9 hours
Mass (Earth = 1)	318	95.18	14.5	17.14	0.0022
Density (water = 1)	1.33	0.69	1.29	1.64	2.03
Temperature (°F on the surface)	-292°	-292°	-346°	-346°	-364°
Number of moons	16	18	15	8	1

THE SUN IN FIGURES

Mass (kg)	$1,989 \times 10^{30}$
Mass (Earth = 1)	332.830
Equatorial radius (miles)	431,853
Average density (g/cm³)	1.410
Distance to the Earth (miles)	93,205,679
Distance to the nearest star (miles)	2.5×10^{10}
Period of rotation (days)	25–36
Translation velocity	240 million years to orbit the galaxy
Average surface temperature	10,832°F
Core temperature	25,200,032°F

Main chemical components

Hydrogen	92.1 %	Iron	0.0037 %
Helium	7.8 %	Silicon	0.0031 %
Oxygen	0.061 %	Magnesium	0.0024 %
Carbon	0.030 %	Sulfur	0.0015 %
Nitrogen	0.0084 %	Others	0.0015 %
Neon	0.0076 %		

THE MOON IN FIGURES

Mass (kg)	$7,349 \times 10^{22}$
Mass (Earth = 1)	0.012298
Equatorial radius (miles)	1,080
Equatorial radius (Earth = 1)	0.27241
Average density (g/cm³)	3.34
Average distance from the Earth (miles)	148,418
Rotational period (days)	27.32166
Orbital period (days)	27.32166
Average orbital velocity (miles/s)	0.64
Surface gravity at the equator (m/s²)	1.62
Surface gravity (Earth = 1)	0.16
Average surface temperature (day)	224.6°F
Average surface temperature (night)	-243.4°F

GLOSSARY

Astronomical unit	This unit of distance is equivalent to 93 million miles (150 million km). It is approximately equal to the average distance from the Earth to the sun, measured from center to center. It is used to compare distances from one object to another within the solar system.
Black holes	Collapsed stars with a very strong gravitational field. No electromagnetic radiation or light can escape from them; this is why they are "black." They are surrounded by a spherical "edge," which allows light to enter, but not to leave.
Cosmic or interstellar dust	This is made up of solid particles of less than a micrometer (one-millionth of a meter) in size, joined with clouds of very low density. Among the elements that make up cosmic dust are hydrogen, carbon, and, in much lower quantities, silicates, and other compounds, such as organic molecules and water.
Light year	Distance that light covers in one year, at a velocity of 186,000 meters/sec; one light year is equivalent to 5.878×10^{12} miles, or 63.240 astronomical units (AU).
Nebulae	These are clouds of interstellar gas. They may appear dark if their temperature is very low, or luminous (emitting light) when their temperature is high enough to excite hydrogen atoms, their main gaseous component. Besides interstellar gases such as hydrogen and helium, nebulae also contain cosmic dust.
Neutron stars	These stars are believed to be the remains of stars that have gone through a supernova phase. A neutron star is an object with a mass between 1.5 and 3 times the mass of the sun, but with a radius of only 6.2 miles (10 km). These stars are essentially made up of neutrons and are the end result of the evolution of a star with great mass.
Novas and supernovas (variable stars)	Novas and supernovas have a certain similarity to each other. Both are stars that show significant and unpredictable variations in brightness, but these are actually two very different phenomena. A nova increases in brightness by a factor of 10, and then decreases to a level of brightness very close to what it showed before the outburst. The star is not destroyed, and may repeat the process thousands of years later. Supernovas, on the other hand, are stars that die in a catastrophic manner.
Pulsars	Objects that give off signals in a pulsating manner. The signals are separated with great precision, with periods between a few milliseconds to a few seconds. They show the existence of stars in an initial state of formation, made up only of neutrons—atomic particles—and not of complete atoms.
Quasars	The word *quasar* is an acronym for "quasi-stellar radio source." They are distant objects that give off large amounts of energy, with radiation similar to that of stars. Their image is similar to that of a common star, although they are 1,000 times brighter than an average-sized galaxy.
Red giants	When the intermediate layer of a star consumes all the hydrogen of its core and fuses it into helium, it puts pressure from the inside toward the outside, pushing against the outer layers of the star. These outer layers begin to expand and then cool, so that the star reaches very large dimensions (giant) and very low temperatures (red).
White dwarf	White dwarfs represent the final phase of evolution for stars with less than 1.4 solar mass. When the reserves of nuclear fuel are exhausted, only a collapse can keep producing energy. In this way, the star cools and becomes more compact.

INDEX